Dangerous Bugs

MOSQUITOES

MEGAN COOLEY PETERSON

BLACK RABBIT BOOKS

Bolt is published by Black Rabbit Books
P.O. Box 227, Mankato, Minnesota, 56002
www.blackrabbitbooks.com
Copyright © 2024 Black Rabbit Books

Alissa Thielges, editor; Michael Sellner, designer
and photo researcher

Library of Congress Cataloging-in-Publication Data
Names: Peterson, Megan Cooley, author.
Title: Mosquitoes / by Megan Cooley Peterson.
Description: Mankato, Minnesota: Black Rabbit Books, [2024] |
Series: Bolt: Dangerous bugs | Includes bibliographical references and index. |
Audience: Ages 8–12 | Audience: Grades 4–6 |
Summary: "Able to deliver deadly diseases, mosquitoes are more than just a pest.
Get up close to these dangerous bugs through gross photos, leveled text, and
engaging infographics that'll make readers squirm"—Provided by publisher.
Identifiers: LCCN 2022024131 (print) | LCCN 2022024132 (ebook) |
ISBN 9781623105808 (library binding) | ISBN 9781623105860 (ebook)
Subjects: LCSH: Mosquitoes—Juvenile literature.
Classification: LCC QL536 .P466 2024 (print) | LCC QL536 (ebook) |
DDC 595.77/2—dc23/eng/20220624
LC record available at https://lccn.loc.gov/2022024131
LC ebook record available at https://lccn.loc.gov/2022024132

Printed in China

Image Credits
Alamy: blickwinkel 22–23,
Razvan Cornel Constantin cover; Dreamstime: Anat Chantrakool 31,
Konstantin Nechaev 14, Lee Hua ming 22–23,
Razvan Cornel Constantin 13, Sucharat Chounyoo 28–29; Minden Pictures: Mark Moffett 28–29;
Science Source: Dr. Tony Brain 3, Eye of Science 14,
22–23, IRD/Vectopole Sud/PATRICK LANDMANN
28–29, TIM VERNON 25; Shutterstock: asturfauna
22–23, Boguslaw Szczepanski 1, 8–9, Cornel Constantin 32, FamVeld 26–27, ivSky 17, Kletr 10–11,
Kwangmoozaa 25, 28–29, Lamnoi Manas 18,
Michael Shilyaev 4–5, nechaevkon 6, 8–9,
praditkhorn somboonsa 17, Radovan1 27,
Thammanoon Khamchalee 21, thatmacroguy 21, Zhenyakot 7

CONTENTS

CHAPTER 1
A Bloodthirsty Bug....4

CHAPTER 2
Size and Features.....8

CHAPTER 3
Where They Live and
What They Eat........16

CHAPTER 4
Life Cycle............20

CHAPTER 5
Watch Out!.........24

Other Resources...........30

A BLOODTHIRSTY Bug

A mosquito flies around a swampy area as the sun sets. The bug smells **carbon dioxide**. It follows the smell to a group of people. The mosquito buzzes toward them. It is almost feeding time.

Time to Feast

The mosquito lands on a person's bare arm. It jabs its mouthpart into the person's skin. Then the bug fills up on blood. The person swats the pest away. But it is too late. As the mosquito ate, it spread a deadly disease.

In the United States, only 12 types of mosquitoes can make people sick.

CHAPTER 2

Size and FEATURES

Mosquitoes might be small. But they can be big pests! Their tiny bodies have three main parts and six legs. Two wings grow from the thorax. Scales cover these wings.

How Big Is a Mosquito?

LENGTH
0.125 to 0.75 INCH
(0.32 to 2 centimeter)

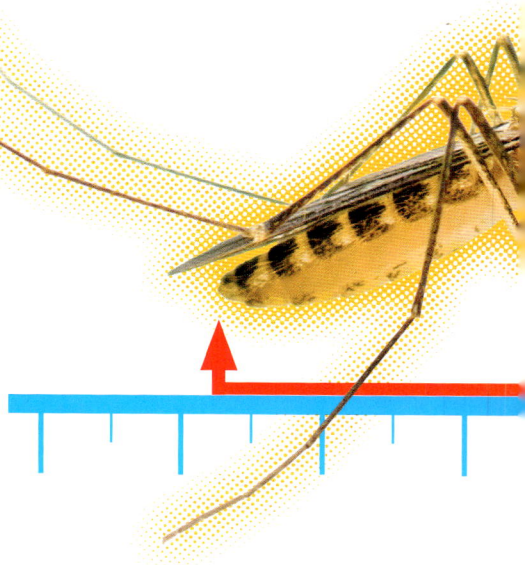

WEIGHT less than **1/1,000** **OUNCE** (2.5 milligram)

PARTS OF A MOSQUITO

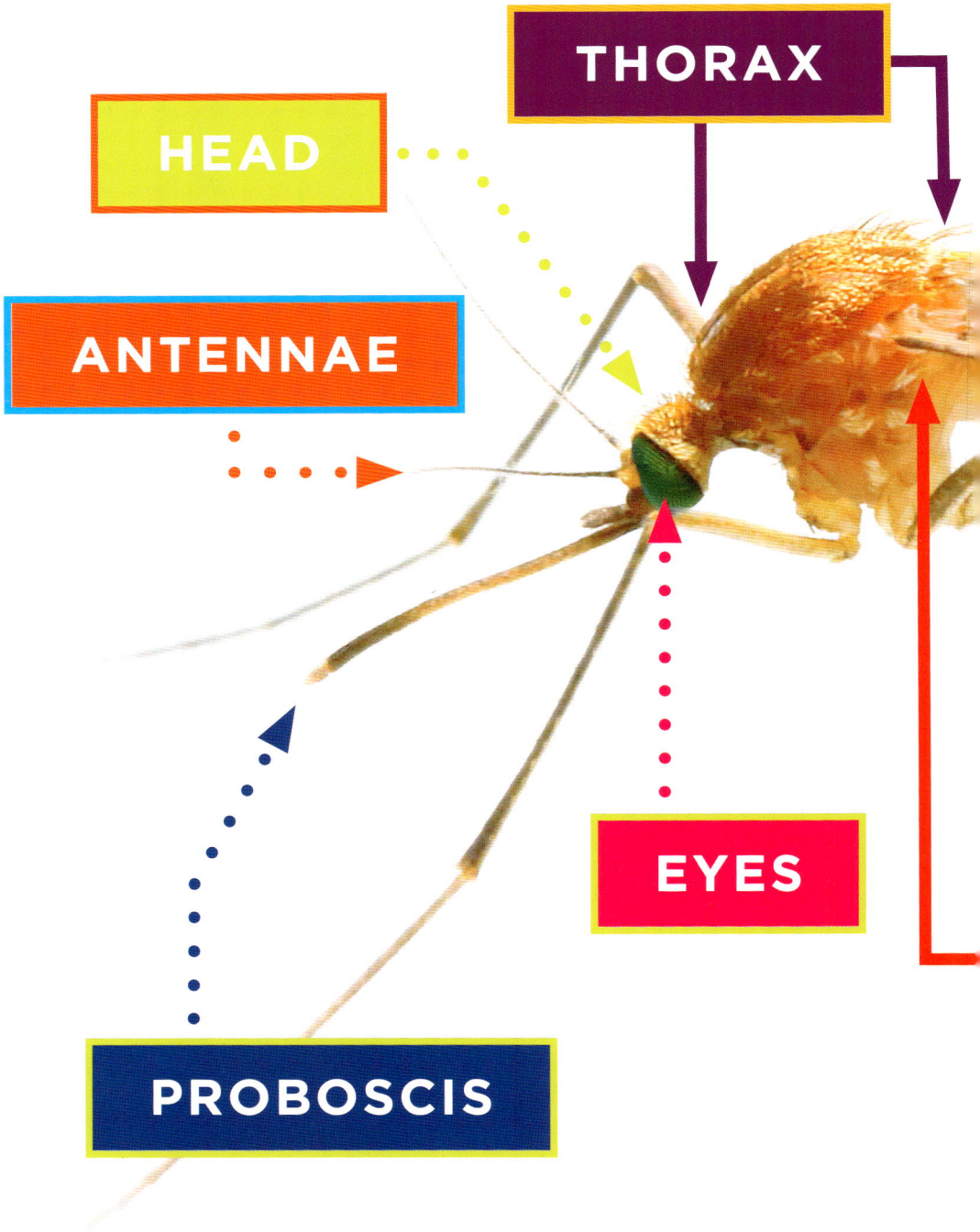

THORAX

HEAD

ANTENNAE

EYES

PROBOSCIS

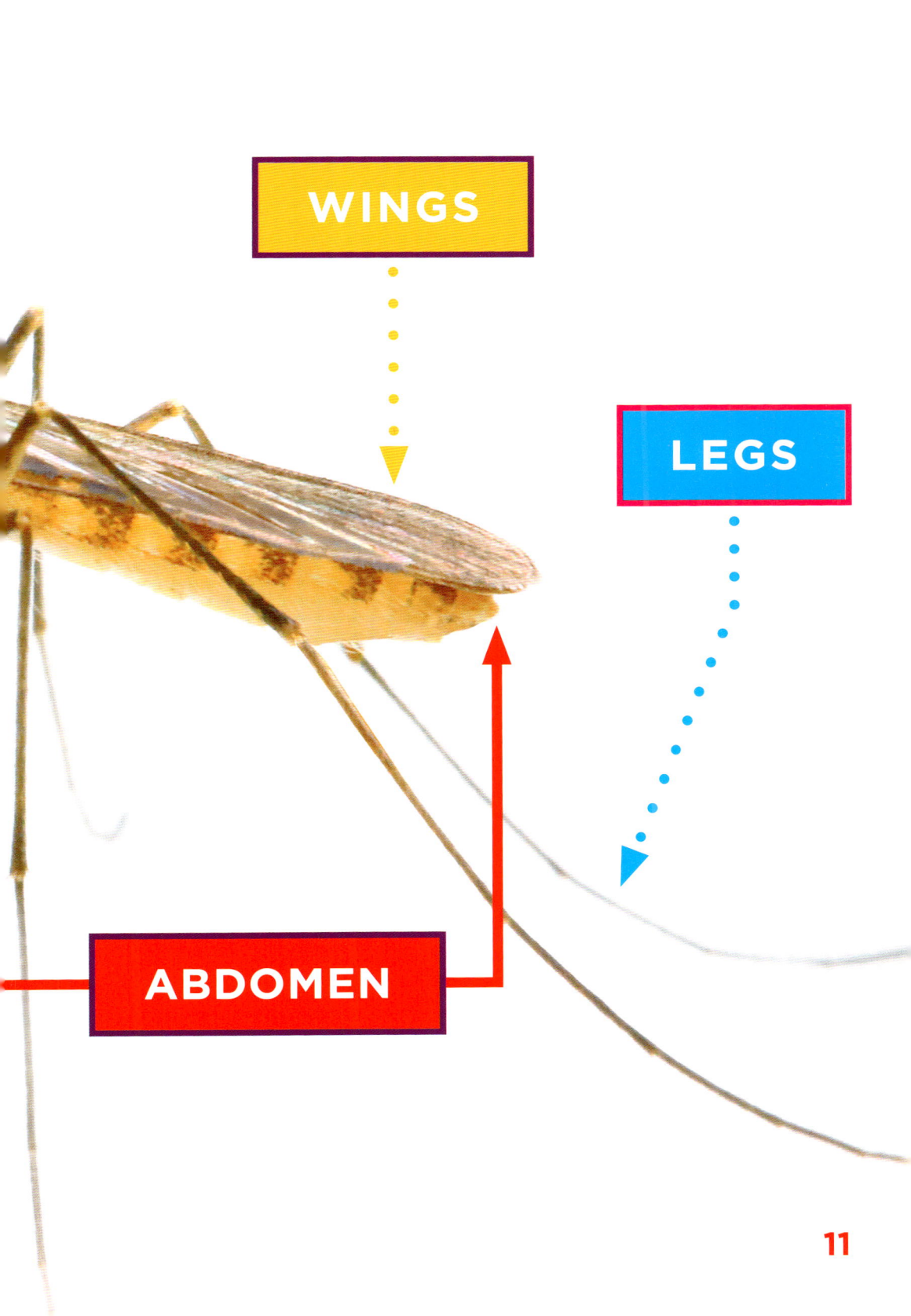

WINGS

LEGS

ABDOMEN

11

Super Senses

A mosquito's senses help it find food. These bugs have two compound eyes. Each eye has lots of small lenses. They can see many directions at once and sense movements. Mosquitoes smell with their antennae. Two **organs**, called palps, also help them sense smells.

Mosquitoes can smell carbon dioxide from more than 100 feet (30 meters) away.

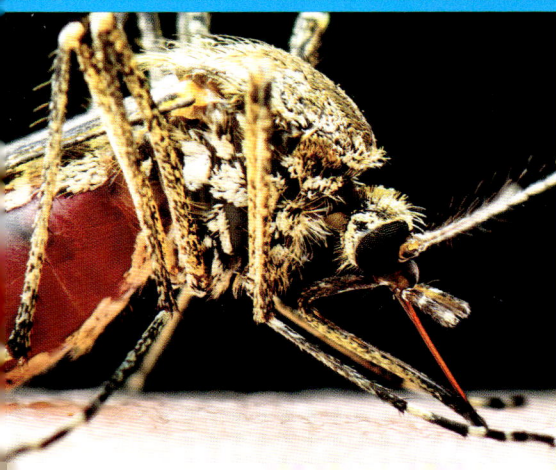

Mosquito saliva keeps its prey from feeling the bite. It also keeps the blood flowing.

Slurp!

Chomp! Mosquitoes bite with a proboscis. The proboscis has six needlelike parts called stylets. When a mosquito bites, the stylets pierce the skin. The mosquito sucks its meal through its mouthpart. As the bug sucks, its body swells with blood.

CHAPTER 3

Where They Live and

WHAT THEY EAT

Mosquitoes live almost everywhere. They like warm, wet weather. Their eggs need calm waters to hatch. Mosquitoes zoom around grassy areas and in forests. They stay close to ponds, swamps, and other areas with standing water.

Swamps

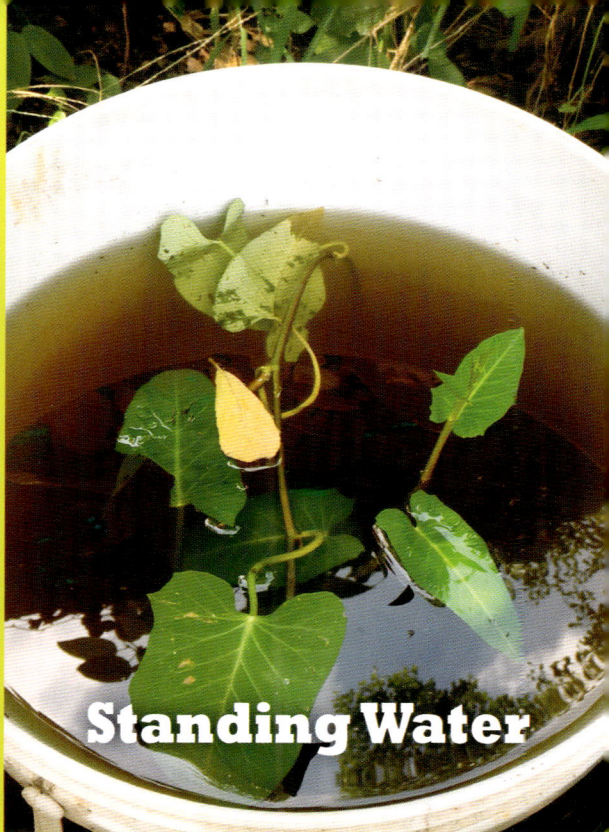

Standing Water

Where Mosquitoes Live

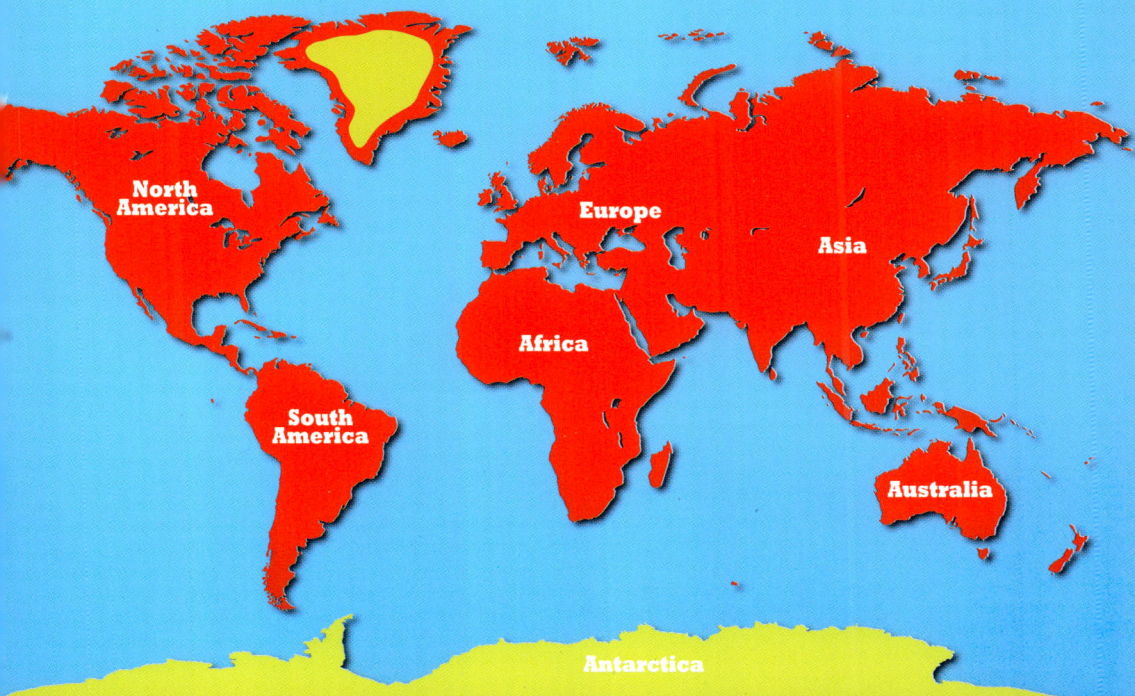

North America

Europe

Asia

Africa

South America

Australia

Antarctica

Mealtime

Mosquitoes don't always eat blood. They drink **nectar** from flowers. Nectar has sugar. Sugar gives mosquitoes energy. Females eat blood when they need to lay eggs. Most females need **protein** from the blood to make the eggs. Males don't suck blood at all.

CHAPTER 4

LIFE CYCLE

All mosquitoes start life as tiny eggs. Females lay clusters of eggs on a water's surface or in damp soil. Then the females fly away.

After a few days, the young hatch. The larvae live in water. Many swim to the surface for air. They **molt** many times. Finally, they become pupae. Pupae shed their skin to become adults.

EGGS

LARVAE

People call the larvae "wrigglers."
They swim using wriggling movements.

From Egg to ADULT

Females lay eggs. Some eggs don't hatch for many months. Most hatch after only a few days.

EGG

Most mosquitoes live two to four weeks. Females that avoid getting squashed can live up to six months.

LARVA

Larvae molt about four times. They grow into pupae in about five days.

PUPA

The pupal stage lasts for two or three days.

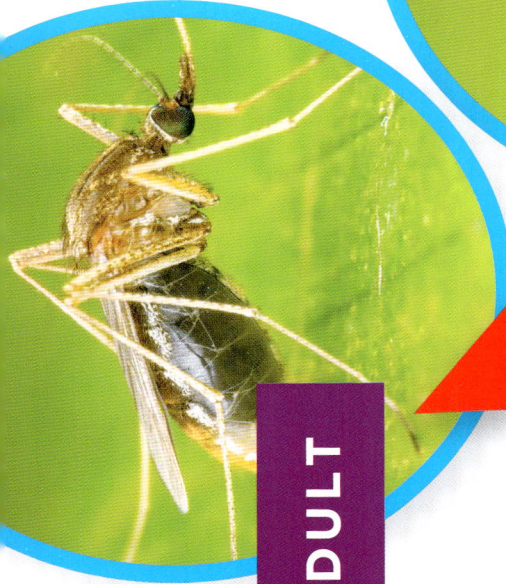

ADULT

CHAPTER 5

WATCH Out!

Mosquitoes are the deadliest animals on Earth. Each year, they kill about 1 million people. These bugs are so deadly because they can carry **parasites** and **viruses**. When they bite people, they spread diseases. Some mosquitoes carry malaria and West Nile virus. These diseases can be deadly to people.

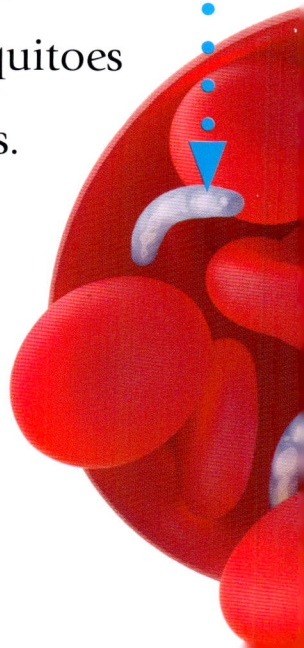

Most people get itchy red bumps from mosquito bites. They're **allergic** to the spit.

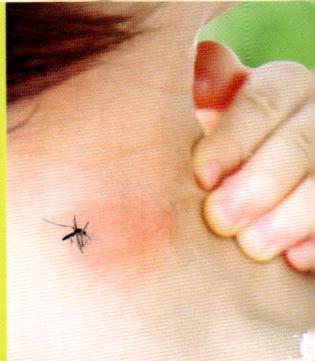

Staying Safe

People can keep mosquitoes outside by sealing doors and windows. When outside, people should wear bug spray. Water should be removed from buckets or pools too. If not removed, females will lay eggs there. People should also stay inside at dawn and dusk. Mosquitoes are most active at these times. Avoiding these dangerous bugs is the best way to stay safe.

BY THE NUMBERS

ABOUT 200

types of mosquitoes in North America

up to 200
NUMBER OF EGGS
A FEMALE CAN LAY
AT ONE TIME

200 to 500

WINGBEATS
PER SECOND

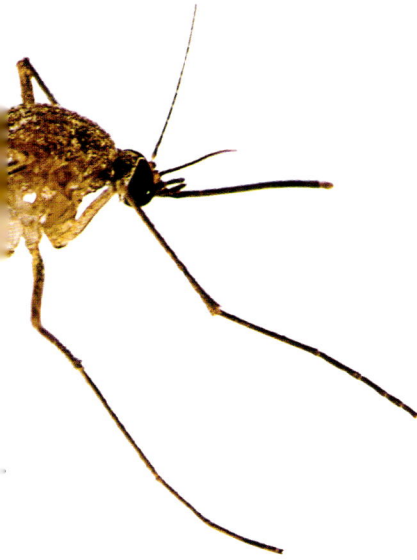

1 to 1.5
MILES
(1.6 to 2.4 kilometers)
PER HOUR
average flying speed

about
3,500

**types of
mosquitoes
in the world**

GLOSSARY

allergic (uh-LUR-jik)—having a medical condition that causes someone to become sick after eating, touching, or breathing something that is harmless to most people

carbon dioxide (KAR-buhn dy-AHK-siyd)—a colorless gas that is formed in the process of breathing

molt (MOLT)—to lose a covering of hair, feathers, or skin and replace it with a new growth

nectar (NEK-tuhr)—a sweet liquid made by plants and flowers

organ (OHR-guhn)—a structure inside the body made of cells and tissues that performs a specific function

parasite (PAR-uh-syt)—a plant or animal that lives in or on another plant or animal and causes harm

protein (PROH-teen)—a small substance in plant or animal cells

BOOKS

Huddleston, Emma. *Tiger Mosquitoes.* Invasive Species. Lake Elmo, MN: Focus Readers, 2022.

Kaiser, Brianna. *Malaria: An Ongoing Threat.* Deadly Diseases. Minneapolis: Lerner Publications, 2022.

O'Brien, Cynthia. *The War Against Malaria.* Wars Waged Under the Microscope. New York: Crabtree Publishing Company, 2022.

WEBSITES

Mosquito
kids.nationalgeographic.com/animals/invertebrates/facts/mosquito/

Mosquitoes
www.dkfindout.com/uk/animals-and-nature/insects/mosquitoes/

Mosquito Facts for Kids
pestworldforkids.org/pest-guide/mosquitoes/

INDEX

B

bites 14, 15, 24, 25

D

diseases 7, 24

E

eggs 16, 19, 20, 22, 27, 28

F

features 7, 8, 10–11, 15

feeding 4, 7, 15, 19

H

habitats 16, 17

L

life cycle 20, 22–23

R

range 17

S

safety 27

senses 12

size 8–9

speed 29

T

types 28, 29